BASIC CHEMISTRY (2)

Preface

This book is genuinely written for grasping the fundamental concept of chemistry. It is aimed to the secondary level students. It can serve as a reference for a particular topic. It is also useful for various competitions.

Index

Introduction

Necessity is the mother of invention. Human curiosity and imagination had added a wing to our flight of invention. Science has no destination but journey. As a domain of science chemistry also deals with what? Why? How? From what the matter around us made up of? And what are the properties of these matters? These are the subjects of chemistry. It has a golden history from alchemists to the detection of higgs boson (commonly referred by journalists "the god particle"). It is assumed that alchemists used to convert other elements into gold. Now we are observing the same thing only the difference between two elements is the number of fundamental particles electron, proton, and neutrons. Now, we are aware of that the whole universe is made up of 118 elements (discovered till now). Our body, the daily items we use, medicine we take, and our surroundings all are the result of these elements only.

It is the second book in the series of basic chemistry. First book deals with some elementary concepts like from what the matter around us made up of? Why some substances react and some not? In this book we will try to explore, actually what happens when these substances react? What are the factors which govern these reactions? We shall also perceive the properties of these substances. For our convenience we will categories the substances as acid, base, salt, metal, non-metal and carbon compounds. In the last chapter our emphasis will be at the classification of elements.

CHAPTER 5

CHEMICAL REACTIONS AND EQUATIONS

INTRODUCTION

In our day to day life we come across many reactions. Rusting of iron articles, formation of curd from milk and digestion of food in our body all are the examples of chemical reactions. For every reaction there are some common features that are observed. In a chemical reaction one or more following properties must be seen. These are:

a. Change in state
b. Change in colour
c. Evolution of gases
d. Change in temperature

Now, it is easily observable that during the rusting of iron like rime of our bicycle its colour changes to brown. it is the change of colour. During the formation of curd liquid changes to solid. In the digestion of food almost all changes like change of state, change of colour, change of temperature and evolution of gases take place.

Chemical reaction and equation

Chemical reaction is an actual process while chemical equation is the symbolic representation of chemical reaction.

In a chemical equation the substances which take part are known as reactants and the substances which are formed referred as the product. In an ideal chemical equation the reactants are kept left side of arrow and added to each other by plus (+) sign and products are kept right side of the arrow.

Example 1:

$$H_2 + O_2 \rightarrow H_2O$$

reactants   products

In a chemical equation the state of the reactants and products are also shown i.e.

Solid (s), liquid (l), gas (g), aqueous (aq), precipitate (ppt).

Balanced chemical equation

In a balanced chemical equation the number of atom of each element is equal in reactant sides as well as product side.

Example 2:

$$CH_4 + 2O_2 \rightarrow CO_2 + 2H_2O$$

C=1		C=1
H=4	=	H=4
O=4		O=4

If a given reaction is not balanced it is called skeletal equation.

Balancing a chemical equation

Method 1: giving priority to oxygen and hydrogen

For balancing a chemical equation if we give priority to oxygen and then hydrogen it becomes quite easier to balance.

Example 3: balance the following reaction

$Fe + O_2 \rightarrow Fe_2O_3$

Here we observe that in LHS there are two atoms of oxygen and 3 atoms of oxygen in RHS. If we multiply LHS by 3 and RHS by 2 the oxygen becomes balanced.

$Fe + 3O_2 \rightarrow 2Fe_2O_3$

Now,

We see that oxygen is balanced then for Fe it is 1 in LHS and 4 in RHS therefore multiply by 4 in LHS and reaction becomes balanced.

$4Fe + 3O_2 \rightarrow 2Fe_2O_3$

Method 2: an algebraic method

If you are quite interested in mathematics. Here is a handy method for you.

Example 4: balance the following equation

$Fe + O_2 \rightarrow Fe_2O_3$

Step1:

Let the coefficient of Fe, O_2 and Fe_2O_3 be a, b and c respectively. Then equation becomes

$aFe + bO_2 \rightarrow cFe_2O_3$

step2:

let it is balanced equation then

$a = 2c$ and $2b = 3c$

here c has the maximum multipliers so let c=1.

On solving it we get

A=2

B=1.5

C=1

Then equation becomes

$2Fe + 1.5O_2 \rightarrow Fe_2O_3$

Step3:

For getting integral ceeffficients we can multiply by 2 in LHS and RHS.

We get

$4Fe + 3O_2 \rightarrow 2Fe_2O_3$

Note: in many cases step 3 doesn't require.

Types of chemical reactions

1.Combination reaction: when two or more than two reactants combineto form a single product, it is known as combination reaction.

Example:

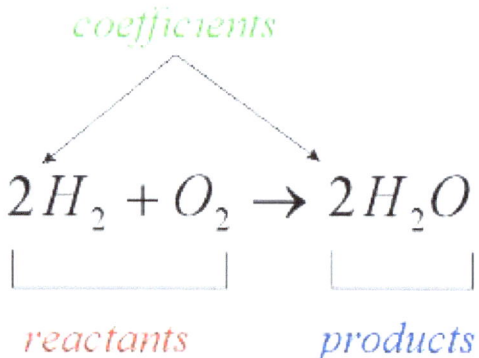

2. Exothermic reaction: the reaction in which heat is released along with the formation of product is know n as exothermic reaction.

Respiration and digestion of food are bestr example of this

Cellular Respiration

$$C_6H_{12}O_6 + 6O_2 \longrightarrow 6CO_2 + 6H_2O + Energy$$

$$CH_4 + 2O_2 \rightarrow CO_2 + 2H_2O + Heat$$
$$C + O_2 \rightarrow CO_2 + Heat$$
$$2H_2 + O_2 \rightarrow 2H_2O + Heat$$

3. Endothermic reaction: Endothermic reaction is the reverse process of exothermic. In this reaction the heat is absorbed along the formation of the products.

$$N_{2(g)} + O_{2(g)} + 180.5kJ \rightarrow 2NO_{(g)}$$

$$2HgO_{(g)} + 180kJ \rightarrow 2Hg_{(l)} + O_{2(g)}$$

4. Decomposition reaction: in decomposition reaction a single reactant breaks into two or more than two products. This reaction can be carried out using heat or electricity and respectievly reffred as thermal or electrical decomposition reaction.

$$NH_4NO_3 \rightarrow N_2O + 2H_2O$$

NH_4NO_3 is Ammonium Nitrate

N_2O is Dinitrogen monoxide

H_2O is water

5. Displacement reaction: this reaction is genuinely reffers to the exchange of ions in when an ionic compound reacts with another higher reactive metal. In actual practice the higher reactive element displaces the lower from its ion.

Single Replacement

a single element replaces a second element in a compound

$$A + BX \rightarrow B + AX$$

Cation Replacement

$$Zn + CuCl_2 \rightarrow ZnCl_2 + Cu$$

Anion Replacement

$$Br_2 + 2KI \xrightarrow{} 2KBr + I_2$$

6. Double Displacement reaction: it is similar to the displacement reaction only difference is that in this reaction both reactants are ionic compounds. In this reaction the ions are exchanged.

Double replacement

$$AB + CD \longrightarrow AD + CB$$

Example: Removal of poisonous barium

$$BaCl_2 + MgSO_4 \longrightarrow BaSO_4 + MgCl_2$$

$$Ba^{2+}_{(aq)} + 2Cl^-_{(aq)} + Mg^{2+}_{(aq)} + SO_4^{2-}_{(aq)} \rightarrow BaSO_{4(s)} + Mg^{2+}_{(aq)} + 2Cl^-_{(aq)}$$

7. Precipitation reaction: in this type of reaction the precipetate is formed which is setteled down to the bottom. This reaction can be also identified in the balanced chemical equation by noticing a downward arrow () .

Na_2SO_4 (aq) + $BaCl_2$ (aq) → $BaSO_4$ (ppt) + 2NaCl (aq)

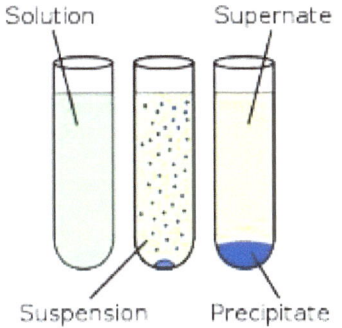

8. Oxidation and Reduction reaction: in a general way oxidation refers to the addition of oxygen or loss of hydrogen and reduction refers to the loss of oxygen or gain of hydrogen. Actually addition of oxygen removes electron and loss of oxygen refers to the gaining of electron.

In a simple way we can state that in a particular reaction if any atom gainig electron it is redced or if it is removing electron it is oxidising. Oxidation and reduction are the supplementry to each other. Both of these processes are take place in single reaction, because if there is removal of electrons certainly removed electron is added to other atom participating in the reaction.

Example:

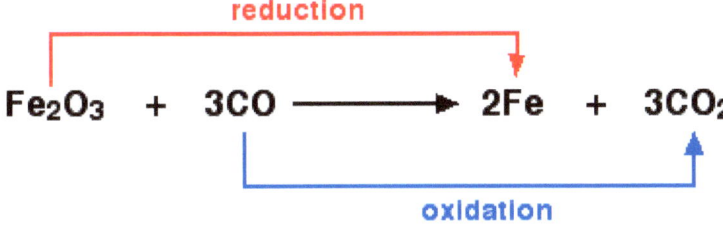

Reactions Affecting Our Life:

Corroision

When any metal comes into the contact of air, water, moistur and acid, it corrodes and this process is known as the corroision. It is very harmful for our countries economy. In a single year due to corroision many thing made up of iron and other metals corrode and government has to repair it. For example the government buses, monument's safety bar made up of iron. Our door, the rim of motor vehicles these are also affected.

Corroision can be minimised using oil, greese on the surface of mettalic part, it prevents the air, water, moist and acid to come into the contact of the metal surface. Galvanisatioin is the another process to prevent it. In this process a layer of zinc is added to the surface of other metals like iron. As zinc is less prone to corrosion, it is a suitable choice. This done using electrolysis processes.

Rancidity

When oily substances like chips comes into the contact of oxygen get oxydised and their tastes changes. This process is refferd as the rancidity. Avoiding this chips packets are flushed with nitrogen gases.

Review Questions

Q1. When Zinc reacts with HCl which gas is released?

    a. H_2

    b. He

    c. O

    d. Cl

Q2. In equation the given reaction, which is reactant?

$H_2 + O2 \rightarrow H2O$

    a. H_2

    b. O2

    c. Both

    d. None

Q3. $Fe + H_2O \rightarrow ? + H_2$

Which compound should be at the place of ? in the above equation?

    a. Fe_2O_3

    b. FeO

    c. Fe_3O_4

    d. $Fe(OH)_3$

Q.4 $3Fe + 4H_2O \rightarrow ? + 4H_2$

Which compound should be at the place of ? in the above equation?

    a. Fe_2O_3

    b. FeO

    c. Fe_3O_4

    d. $Fe(OH)_3$

Q5. $C + O_2 \rightarrow CO_2$ is ?

    a. Displacement reaction

b. Endothermic reaction

c. Exothermic reaction

d. None

Q6. What is the formula of lime stone?

a. CaO

b. $CaCO_2$

c. $Ca(OH)_2$

d. $CaCO_3$

Q7. what is the formua of quick lime?

a. CaO

b. CaCo2

c. Ca(OH)2

d. CaCO3

Q.8 what should be the value of X so that equation becomes balanced?

$2Pb(NO3)2(s) \rightarrow 2PbO(s) + XNO2(g) + O2(g)$

a. 1

b. 2

c. 3

d. 4

Q.9 Fe(s) + CuSO4(aq) → FeSO4(aq) + Cu(s)

Which type of reaction is this?

a. Displacement reaction

b. combination reaction

c. decomposition reaction

d. None

Q.10 which is oxidised in the given reaction?

$2Cu + O_2 \rightarrow CuO$

    a. Cu

    b. O2

    c. CuO

    d. None

Q.11 which is essential for corroision?

    a. air

    b. water

    c. moisture

    d. all of these

Q.12 silver articles become black when exposed to air. In this process which compound is formed?

    a. Silver oxide

    b. Silver nitrate

    c. Silver sulphide

    d. Silver carbonate

Q13. Food kept in air tight jar prevent it from?

    a. oxidation

    b. reduction

    c. corroision

    d. None

Q14. Chips packets are flushed with?

    a. N

    b. O_2

    c. Cl_2

    d. He

Q.15 $2Cu + O_2 \rightarrow 2CuO$

Which type of reaction is this?

    a. Displacement reaction

    b. combination reaction

    c. decomposition reaction

    d. None

Q16. In which type of chemical reaction ions are interchanged?

    a. Displacement reaction

    b. combination reaction

    c. decomposition reaction

    d. double displpacement reaction

Q17. Opposite of combination reaction is?

    a. Displacement reaction

    b. combination reaction

    c. decomposition reaction

    d. double displpacement reaction

Q18. Oxidation is the gain of?

    a. oxygen

    b. electron

    c. protron

    d. none of these

Q19. Reduction is the gain of?

    a. oxygen

    b. electron

    c. protron

    d. hydrogen

Q20. Precipitates are?

    a. Soluble in water

b. Insolu ble in water

c. Partially soluble

d. Float in water

Math the followings

Q21.

A	B
$AgNO_3$	combination reaction
$2H_2 + O_2 \rightarrow 2H_2 + O_2$	black
rancidity	unbalanced reaction
$Fe_2O_3 + Al \rightarrow Al_2O_3 + Fe$	nitrogen

Q22.

A	B
oxidation	gain of electron
reduction	loss of electron
corroision	chips
rancidity	iron

Q23.

A	B
Black and white photography	thermite reaction
Exothermic reaction	AgBr
$CuSO_4$	Au
Noble metal	blue

Q24. What is a balanced chemical equation?

Q.25 what is the difference between combination reaction and decomposition reaction?

Q26. what is the difference between exothermic reaction and endothermic reaction?

Q27. what is the difference between displacement reaction and double displacement reaction?

Q28. What are essential conditions for corroision?

Q29. Why chips packets are flushed with nitrogen gases?

Q30. Write down the reaction which is generally used in black and white photography?

Q31. What is rancidity? Explain with examplpes.

Q32. Why we apply paints on metal articles?

Q33. A shiny brown coloured element 'X' on heating in air becomes black in colour. Name the element 'X' and the black coloured compound formed.

Long answer type questions

Q34. Translate the following statements into chemical equations and then balance them.

(a) Hydrogen gas combines with nitrogen to form ammonia.

(b) Hydrogen sulphide gas burns in air to give water and sulpur dioxide.

(c) Barium chloride reacts with aluminium sulphate to give aluminium chloride and a precipitate of barium sulphate.

(d) Potassium metal reacts with water to give potassium hydroxide and hydrogen gas.

Q35. Balance the following chemical equations.

(a) $HNO_3 + Ca(OH)_2 \rightarrow Ca(NO_3)_2 + H_2O$

(b) $NaOH + H_2SO_4 \rightarrow Na_2SO_4 + H_2O$

(c) $NaCl + AgNO_3 \rightarrow AgCl + NaNO_3$

(d) $BaCl_2 + H_2SO_4 \rightarrow BaSO_4 + HCl$

Q36. Write the balanced chemical equations for the following reactions.

(a) Calcium hydroxide + Carbon dioxide → Calcium carbonate + Water

(b) Zinc + Silver nitrate → Zinc nitrate + Silver

(c) Aluminium + Copper chloride → Aluminium chloride + Copper

(d) Barium chloride + Potassium sulphate → Barium sulphate + Potassium chloride

Q37. Write the balanced chemical equation for the following and identify the type of reaction in each case.

(a) Potassium bromide(aq) + Barium iodide(aq) → Potassium iodide(aq) + Barium bromide(s)

(b) Zinc carbonate(s) → Zinc oxide(s) + Carbon dioxide(g)

(c) Hydrogen(g) + Chlorine(g) → Hydrogen chloride(g)

(d) Magnesium(s) + Hydrochloric acid(aq) → Magnesium chloride(aq) + Hydrogen(g)

CHAPTER 6

Acids, Bases and Salts

Formation of acids

When non metals reacts with oxygen it produces non mettalic oxide.

$$C_{(s)} + O_{2(g)} \longrightarrow CO_{2(g)}$$

$$N_{2(g)} + O_{2(g)} \longrightarrow 2NO_{(g)}$$

$$S_{8(s)} + 8O_{2(g)} \longrightarrow 8SO_{2(g)}$$

When non mettalic oxide reacts with water acid is formed.

Nonmetal oxide+water $\longrightarrow$ Acid

$$CO_{2(g)} + H_2O_{(l)} \longrightarrow H_2CO_{3(aq)}$$

$$SO_{3(g)} + H_2O_{(l)} \longrightarrow H_2SO_{4(l)}$$

$$P_2O_{5(g)} + 3H_2O_{(l)} \longrightarrow 2H_3PO_{4(aq)}$$

Formation of bases

When metals reacts with oxygen it produces mettalic oxide.

$$2Mg_{(s)} + O_{2(g)} \longrightarrow 2MgO_{(s)}$$

$$4Fe_{(s)} + 3O_{2(g)} \longrightarrow 2Fe_2O_{3(s)}$$

$$2Mg_{(s)} + O_{2(g)} \longrightarrow 2MgO_{(s)}$$

$$4Li_{(s)} + O_{2(g)} \longrightarrow 2Li_2O_{(s)}$$

$$2Cr_{(s)} + 3O_{2(g)} \longrightarrow 2CrO_{3(s)}$$

$$4Cu_{(s)} + O_{2(g)} \longrightarrow 2Cu_2O_{(s)}$$

When mettalic oxide reacts with water base is formed.

$$Metal\ Oxide + Water \longrightarrow Metallic\ Hydroxide$$

$$MgO_{(s)} + H_2O_{(l)} \longrightarrow Mg(OH)_{2(s)}$$

$$CaO_{(s)} + H_2O_{(l)} \longrightarrow Ca(OH)_{2(s)}$$

$$ZnO_{(s)} + H_2O_{(l)} \longrightarrow Zn(OH)_{2(s)}$$

In a general way we can understand the formation of base by a very simple activity.

Activity 1:

Step 1: take a magnesium ribbon and clean its upper surface by sand paper.

Step2: burn it with the help of a pair of tongs and a spirit lamp.

Step 3: collect its ash and dissolve in a glass of water.

Now you have a base in your glass.

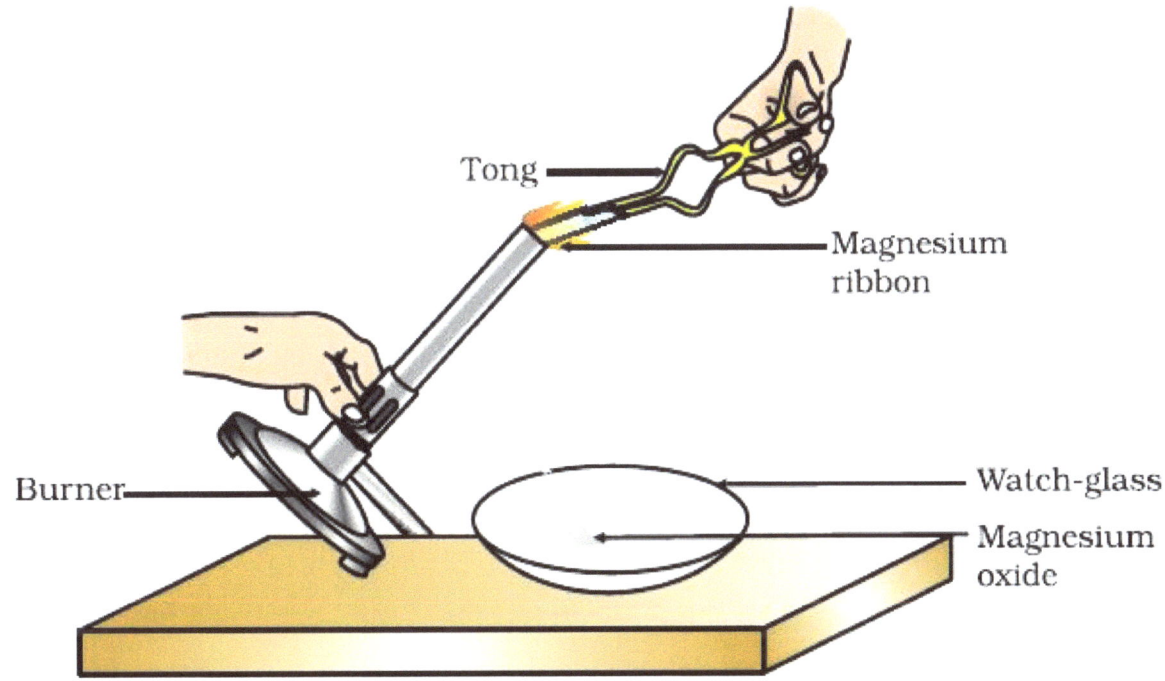

Burning of a magnesium ribbon in air and collection of magnesium oxide in a watch-glass

Formation of salts

When acid reacts with base, both neutralize each other and this reaction is referred as neutralization reaction.

The product of this reaction is known as salt.

Example:

Acid   Base   Water   Salt

$$HCl + NaOH \rightarrow H_2O + NaCl$$

Now we have a general idea of formation of acids, bases and salt. In the next section we will discuss some physical and chemical properties of acid, base and salt.

Detection of acids and bases in general life

Generally acids are sour in taste while bases are bitter in nature. Acid turns blue litmus paper to red while base changes red to blue. When acids and bases are in solution or aqueous form we can check whether they are acids or bases using many natural flowers and petals. These flowers or petals are referred as natural indicator. Some of these are red cabbage, turmeric, colored petals of flowers such as hydrangea, petunia and geranium.

There are many synthesized indicators. These are shown in the following table.

Indicator	pH range over which colour change occurs	colour of acid form	colour of conjugate base form
methyl orange	2,1 - 4,4	orange	yellow
methyl red	4,2 - 6,2	red	yellow
bromothymol blue	6,0 - 7,8	yellow	blue
phenolphthalein	8,3 - 10,0	colourless	pink
alizarin yellow	10,1 - 12,1	yellow	red

Chemical properties of Acids and bases:

Reaction with metals

When acids or bases react with metal salt and hydrogen gas is produced.

$NaOH + Zn \rightarrow Na_2ZNO_2 + H_2$

Reaction with metal carbonates and hydrogen carbonates

On reaction with metal carbonate or metal hydrogen carbonate acids and bases produces salt and water with carbon di oxide gas.

Hydrochloric acid + sodium carbonate $\rightarrow$ water + sodium chloride + carbon dioxide

$HCl \quad + \quad Na_2CO_3 \quad \rightarrow \quad H_2O \quad + \quad NaCl \quad + CO_2$

Sulfuric acid + ammonium hydrogen carbonate $\rightarrow$ water + ammonium sulfate + carbon dioxide

$H_2SO_4 \quad + \quad 2NH_4HCO_3 \quad \rightarrow H_2O \quad + \quad (NH_4)_2SO_4 \quad + \quad 2CO_2$

Metallic oxides with acids

As metallic oxide is basic in nature when it reacts with any acid salt is formed with the release of hydrogen gas.

Non metallic oxides with base

As non metallic oxide is acidic in nature when it reacts with any base salt is formed with the release of hydrogen gas.

Acids and bases in aqueous state

When acid or bases are diluted in water they release H^+ or H^- ions respectively. These ions are necessary to give authentic pH test or any indicator test. That's why all test relevant to acids and bases are taken in aqueous state only.

The following table will recapitulate the properties of acids and bases:

Acid + metal	Salt + H_2
Acid + metal hydroxide	Salt + H_2O
Acid + metal oxide	Salt + H_2O
Acid + metal carbonate	Salt + H_2O + CO_2
Acid + metal hydrogen carbonate	Salt + H_2O + CO_2
Acidic oxide + base	Salt + H_2O

pH scale a strong tool

pH scale is a scale designed to test the acidic or basic character of any substance.

In this scale there are 0 to 14 readings. In this reading 7 refers to the neutral substance while less is acidic and more is basic in nature.

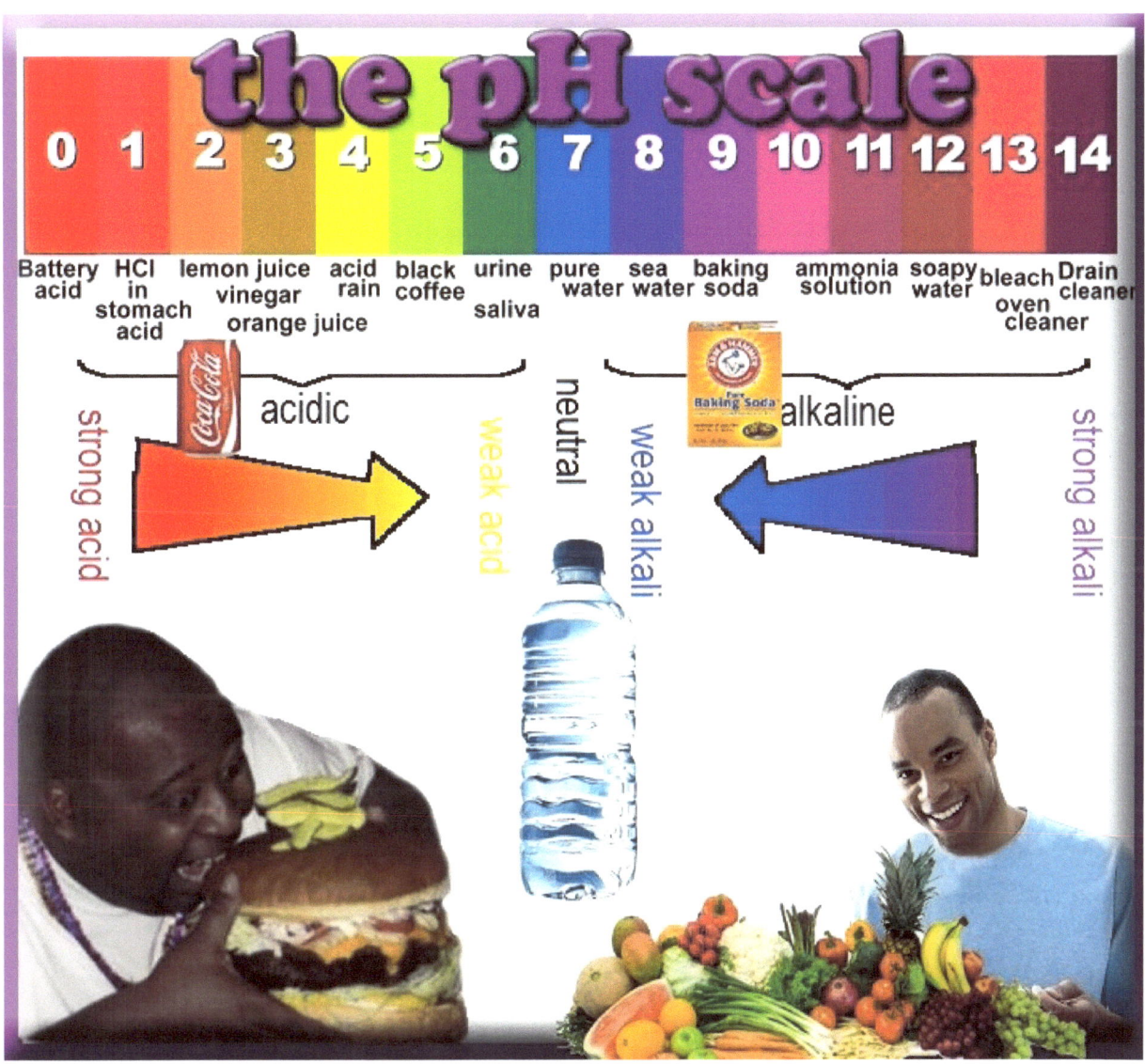

Some important salts:

Bleaching powder

Formation

Chlorine gas is used in the formation of bleaching powder. Bleaching powder is produced by the action of chlorine on dry slaked lime [Ca(OH)2].

$$Ca(OH)_2 + Cl_2 \rightarrow CaOCl_2 + H_2O$$

uses
- (i)   for bleaching cotton and linen in the textile industry
- (ii)   for bleaching wood pulp in paper factories
- (iii)   for bleaching washed clothes in laundry
- (iv)   As an oxidizing agent in many chemical industries
- (v)   For disinfecting drinking water to make it free of germs

Baking soda

Formation

In the formation of baking soda i.e. sodium hydrogen carbonate the basic raw material is sodium chloride.

$$NaCl + H_2O + CO_2 + NH_3 \rightarrow NH_4Cl + NaHCO_3$$

During cooking

$$2NaHCO_3 \rightarrow Na_2CO_3 + H_2O + CO_2$$

Sodium hydrogen carbonate has got various uses in the household.

Uses of sodium hydrogen carbonate (NaHCO3)

(i) For making baking powder, this is a mixture of baking soda (sodium hydrogen carbonate) and a mild edible acid such as tartaric acid. When baking powder is heated or mixed in water, the following reaction takes place –
$NaHCO_3 + H^+ \rightarrow CO_2 + H_2O +$ Sodium salt of acid
Carbon dioxide produced during the reaction causes bread or cake to rise making them soft and spongy.
(ii) Sodium hydrogen carbonate is also an ingredient in antacids. Being alkaline, it neutralizes excess acid in the stomach and provides relief.
(iii) It is also used in soda-acid fire extinguishers.

Washing soda

Formation

When sodium hydrogen carbonate is heated it gives sodium carbonate. On recrystallisation i.e. adding 10 molecules of water to sodium carbonate gives washing soda.

$Na_2CO_3 + 10.H_2O \rightarrow Na_2CO_3.10H_2O$

Uses
(i) Sodium carbonate (washing soda) is used in glass, soap and paper industries.
(ii) It is used in the manufacture of sodium compounds such as borax.
(iii) Sodium carbonate can be used as a cleaning agent for domestic purposes.
(iv) It is used for removing permanent hardness of water.

Plaster of Paris

On heating gypsum at 373 K, it loses water molecules and becomes calcium sulphate hemihydrates. This is called Plaster of Paris, the substance which doctors use as plaster for supporting fractured bones in the right position.

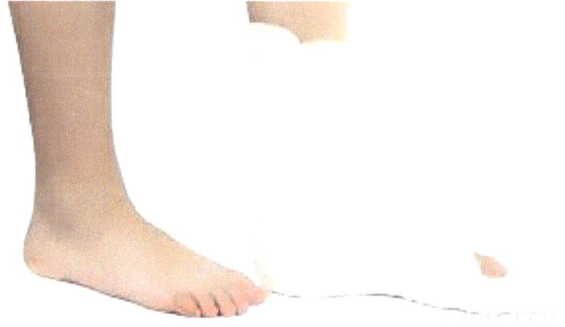

Plaster of Paris is a white powder and on mixing with water, it changes to gypsum once again giving a hard solid mass.

$$CaSO_4 . \frac{1}{2} H_2O + 1\frac{1}{2} H_2O \rightarrow CaSO_4 . 2H_2O$$
(Plaster Of Paris)   (Gypsum)

It is used for making toys, materials for decoration and for making surfaces smooth.

Review questions

Q1. Acid changes blue litmus to?

    a. red

    b. pink

    c. yellow

    d. no change

Q2. Phenolphthalein and methyl oranges are

    a. synthetic indicator

    b. natural indictor

    c. both

    d. none

Q3. In olfactory indicators?

    a. Colour changes

    b. Taste changes

    c. Odour changes

    d. none

Q4. Acid on reaction with metal produces?

    a. H_2

    b. O_2

    c. N_2

    d. None

Q5. Acid on reaction with metal hydrogen carbonate produces?

    a. H_2

    b. CO_2

    c. N_2

    d. None

Q6. In aqueous solution acid produces which type of ions?

a. H$^+$

b. OH$^-$

c. N$_2$

d. None

Q7. The process of dissolving an acid or a base in water is highly?

a. endothermic

b. exothermic

c. cold

d. None

Q8. pH scale measure which ion in a solution?

a. H$^+$

b. OH$^-$

c. Cl$^-$

d. None

Q9. A solution has pH value 7. It is likely to be

a. acid

b. base

c. salt

d. none of these

Q10. Our body works within pH range?

a. 2-4

b. 12-14

c. 6-10

d. 7-7.8

Q11. Milk of magnesia is?

a. acid

b. weak acid

c. antacid

d. None

Q12. Acetic acid is present in?

    a. Apple

    b. guava

    c. vinegar

    d. None

Q13. Oxalic acid is present in?

    a. potato

    b. tomato

    c. onion

    d. none

Q14. pH value of a salt is?

    a. 6

    b. 7

    c. 8

    d. 9

Q15. Formula for bleaching powder is?

    a. $CaOCl_2$

    b. $CaOCl$

    c. $CaCl_3$

    d. None

Q16. Na_2CO_3 is?

    a. Washing soda

    b. Sodium carbonate

    c. Sodium hydrogen carbonate

    d. None

Q17. Which is used for removing permanent hardnessof water?

a. $CaSO_4.1/2H_2O$

b. $CaSO_4.2H_2O$

c. $CaSO_4$

d. Washing soda

Q18. A solution turns blue litmus red its pH is likely to be?

a. 1

b. 7

c. 8

d. 9

Q19. A solution reacts with crushed egg-shells to give a gas that turns lime-water milky. The solution contains

(a) NaCl

(b) HCl

(c) LiCl

(d) KCl

Q20. Which one of the following types of medicines is used for treating indigestion?

(a) Antibiotic

(b) Analgesic

(c) Antacid

(d) Antiseptic

Match the following:

Q21.

A	B
Acetic acid	nettle sting
Citric acid	curd
Lactic acid	vinegar
Methanoic acid	orange

Q22.

A	B
Bleaching powder	$CaSO_4.2H_2O$
Baking soda	$CaSO_4.1/2H_2O$
Washing soda	$CaOCl_2$
Plasterof paris	$Na_2CO_3.10H_2O$
Gypsum	$NaHCO_3$

Q23.

A	B
acid	red cabbage
base	salt
pH 7	H^+
natural indicator	OH^-

short answer type Questions:

Q24. Write down physical diffferences between acid and bases.

Q25. Write down chemical diffferences between acid and bases.

Q26. What are common properties between acid and base?

Q27. What is salt and how it is formed?

Q28. What are anta-acids? How they help us in digestion?

Q29. Write down the uses of baking soda?

Q30. How does plaster of paris is formed?

Q31. Why should curd and sour substances not be kept in brass and copper vessels?

Q32. Why does dry HCl gas not change the colour of the dry litmus paper?

Long answer type questions:

Q33. Suppose you are sent to a planet were total soil is acidic in nature. It is also not suitable for agriculture purpose. You have to grow plant there. What procedure you can use to solve this problem?

Q34. Write down the formation and uses of following compounds:

a. Washing soda
b. Baking soda
c. Bleaching powder
d. Plaster of paris

Q35. Write down five chemical properties of acids. Explai with chemical equation in each case.

Q36. Write down five chemical properties of bases. Explai with chemical equation in each case.

Chapter 7

Metals and Non Metals

Introduction

Physical properties of metals and non metals

Hardness

Hardness refers to strength of a material. Generally metals are hard and non metals are soft and brittle. There are some exceptions, lead is a brittle metal. Some metals are so soft that they can be cut by knife such as sodium, potassium, magnesium.

Malleability

Malleability is the property of being beaten in the form of a thin sheet. Metals can be beaten in the thin sheets. This property helps us to use metals sheets, containers etc.

Ductility

Metals can be drawn into the wires this property is known as ductility. Gold is most ductile element that's why ornaments are not form from the pure gold.

Conduction of heat

The materials which allow heat to pass through them are known as the conductor of heat. Metals are good conductor of heat and non metals are bad.

Conduction of electricity

The materials which allow electricity to pass through them are known as the conductor of electricity. Metals are good conductors and non metals are insulator. Graphite is a non metal but a good conductor of electricity.

Sonority

Sonority refers to the sounds produced when a metal is beaten. Due to this property copper or irons are used as a bell.

Metallic Luster

Metals have a shining surface while non metals have not. This property is referred as metallic luster. Iodine is exception; it is lustrous but non metal.

These properties are summarized in the following table:

PROPERTY	METALS	NON METALS
State of matter	These are usually solid, except mercury, which is a liquid at room temperature. Gallium and Caesium melt below 30ºC. So if room temperature is around 30ºC, they may also be in liquid state	These exist in all the three states. Bromine is the only liquid
Density	They usually have high density, except for sodium, potassium, calcium etc.	Their densities are usually low.
Melting point	They usually have a high melting point except mercury, cesium, gallium, tin, lead.	Their melting points are low.
Hardness	They are usually hard, except mercury, sodium, calcium, potassium, lead etc.	They are usually not hard. But the exception is the non-metal diamond, the hardest substance.
Malleability	They can be beaten into thin sheets.	They are generally brittle.
Ductility	They can be drawn into thin wires, except sodium, potassium, calcium etc.	They cannot be drawn into thin wires.
Ductility	They can be drawn into thin wires, except sodium, potassium, calcium etc.	They cannot be drawn into thin wires.
Conduction of heat	They are good conductors of heat.	They are poor conductors of heat.
Conduction of electricity	They are good conductors of electricity.	They are non-conductors, except for carbon in the

Chemical properties of metals:

In the previous lesson we have seen the reaction of metals and non metals with acids, bases, carbonates, hydrogen carbonates, and oxides. In this section we will see what happens when metals and a non metal react.

Reaction with solution of other salts

When metals react with solution of the other salts displacement reaction occurs. About we have all ready discussed in the chemical reaction part. The following reactivity series will help in finding the more reactive and less reactive elements.

The reactivity series

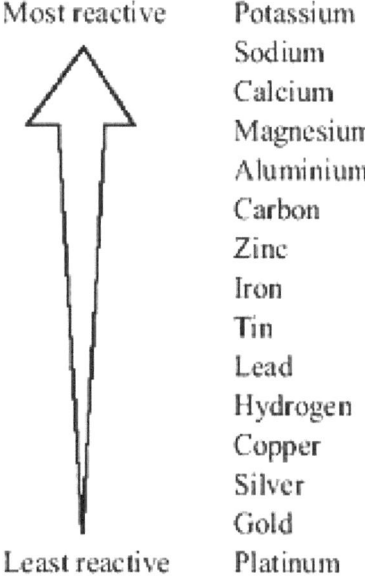

Most reactive — Potassium
Sodium
Calcium
Magnesium
Aluminium
Carbon
Zinc
Iron
Tin
Lead
Hydrogen
Copper
Silver
Gold
Least reactive — Platinum

Reaction with non metals

In the first part of this series (basic chemistry 1), in chapter 0 (some basic concepts) we have studied that when metals and non metals comes into contact electrons are transferred from metal to non metal and anions and cations are formed. These are attracted due to opposite polarity and a strong ionic bond is formed. These compounds are referred as ionic compounds.

Ionic compounds and its properties

(i) *Physical nature*: Ionic compounds are solids and are somewhat hard because of the strong force of attraction between the positive and negative ions. These compounds are generally brittle and break into pieces when pressure is applied.

(ii) *Melting and Boiling points*: Ionic compounds have high melting and boiling points .This is because a considerable amount of energy is required to break the strong inter-ionic attraction.

(iii) *Solubility*: Electrovalent compounds are generally soluble in water and insoluble in solvents such as kerosene, petrol, etc.

(iv) *Conduction* of Electricity: The conduction of electricity through a solution involves the movement of charged particles. A solution of an ionic compound in water contains ions, which move to the opposite electrodes when electricity is passed through the solution. Ionic compounds in the solid state do not conduct electricity because movement of ions in the solid is not possible due to their rigid structure. But ionic compounds conduct electricity in the molten state. This is possible in the molten state since the electrostatic forces of attraction between the oppositely charged ions are overcome due to the heat. Thus, the ions move freely and conduct electricity.

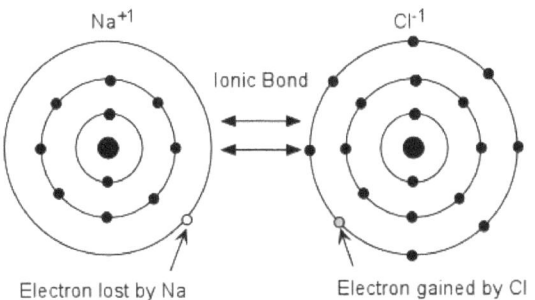

Ores and minerals

Compounds or elements which naturally occur in the earth's crust are known as *minerals*. The best example is water. Profitable minerals are referred as *ores*. If in getting a mineral we are in loss economically it's not an ore. The unwanted sand soil and mud contaminated with ore are referred as gangue.

Extraction of metals

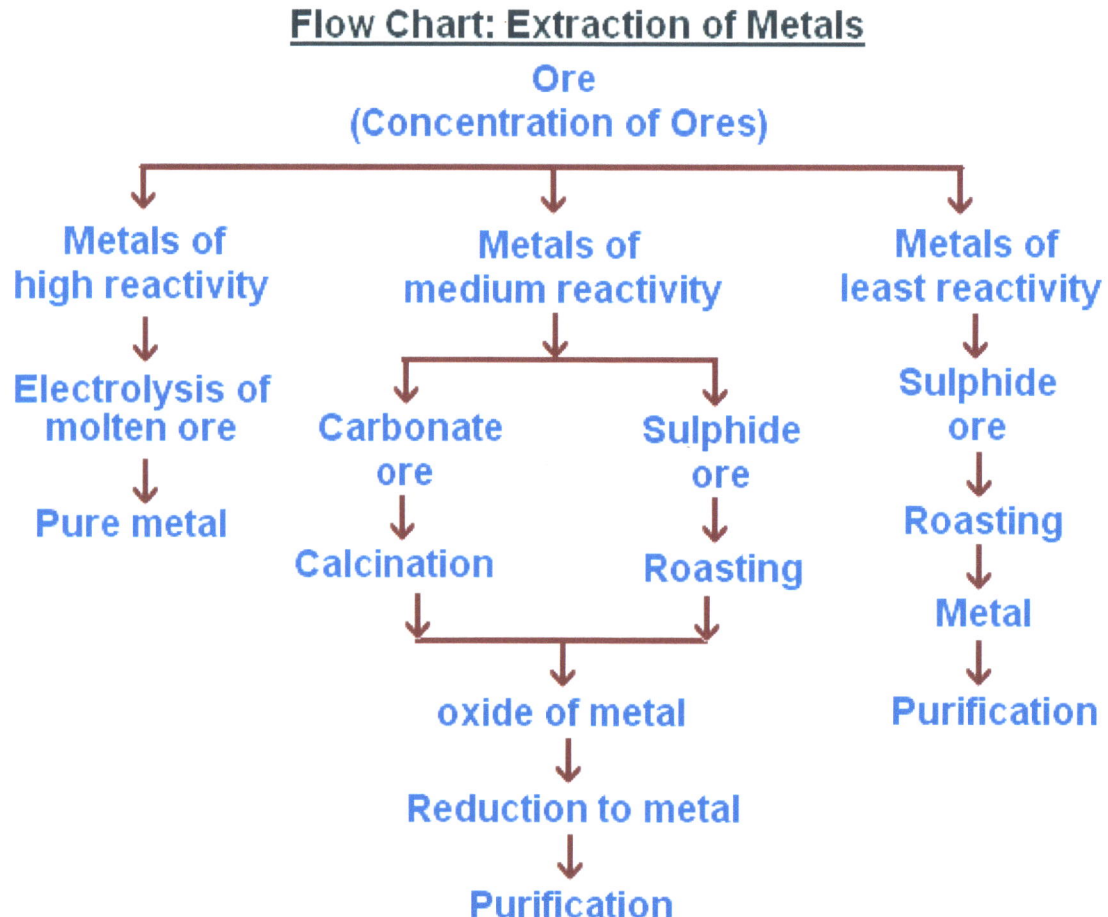

Enrichment of Ores
Ores mined from the earth are usually contaminated with large amounts of impurities such as soil, sand, etc., called *gangue*. The impurities must be removed from the ore prior to the extraction of the metal. The processes several steps are involved in the extraction of pure metal from ores. A summary of these steps is given in flow chart. Each step is explained in detail in the following sections.

The processes used for removing the gangue from the ore are based on the differences between the physical or chemical properties of the gangue and the ore. Different separation techniques are accordingly employed.

Extracting Metals Low in the Activity Series

Metals lows in the activity series are very unreactive. The oxides of these metals can be reduced to metals by heating alone. For example, cinnabar (HgS) is an ore of mercury. When it is heated in air, it is first converted into mercuric oxide (HgO). Mercuric oxide is then reduced to mercury on further heating.

$2HgS(s) + 3O (g) \rightarrow 2HgO(s) + 2SO_2 (g)$
$2HgO(s) \rightarrow 2Hg (l) + O_2 (g)$

Similarly, copper which is found as Cu2S in nature can be obtained from its ore by just heating in air.

$2Cu S + 3O (g) \rightarrow 2Cu O(s) + 2SO (g)$
$2Cu O + 2Cu S \rightarrow 4Cu + 2SO2$

Extracting Metals in the Middle of the Activity Series

The metals in the middle of the activity series such as iron, zinc, lead, copper, etc., are moderately reactive. These are usually present as sulphides or carbonates in nature. It is easier to obtain a metal from its oxide, as compared to its sulphides and carbonates. Therefore, prior to reduction, the metal sulphides and carbonates must be converted into metal oxides. The sulphide ores are converted into oxides by heating strongly in the presence of excess air. This process is known as *roasting*.

The carbonate ores are changed into oxides by heating strongly in limited air. This process is known as *calcination*. The chemical reaction that takes place during roasting and calcination of zinc ores can be shown as follows –

Roasting
$2ZnS(s) + 3O (g) \rightarrow 2ZnO(s) + 2SO2 (g)$

Calcination
$ZnCO_3 \rightarrow (s) ZnO(s) + CO_2 (g)$

The metal oxides are then reduced to the corresponding metals by using suitable reducing agents such as carbon.

For example, when zinc oxide is heated with carbon, it is reduced to metallic zinc.

$ZnO(s) + C(s) \rightarrow Zn(s) + CO (g)$

Besides using carbon (coke) to reduce metal oxides to metals, sometimes displacement reactions can also be used. The highly reactive metals such as sodium, calcium, aluminium, etc., are used as reducing agents because they can displace metals of lower reactivity from their compounds. For example, when manganese dioxide is heated with aluminium powder, the following reaction takes place –

$3MnO_2(s) + 4Al(s) \rightarrow 3Mn (l) + 2Al_2O_3(s) + Heat$

These displacement reactions are highly exothermic. The amount of heat evolved is so large that the metals are produced in the molten state. In fact, the reaction of iron (III) oxide (Fe2O3) with aluminium is used to join railway tracks or cracked machine parts. This reaction is known as the *thermit reaction*.

$Fe_2O_3(s) + 2Al(s) \rightarrow 2Fe (l) + Al_2O_3(s) + Heat$

Extracting Metals towards the Top of the Activity Series
The metals high up in the reactivity series are very reactive. They cannot be obtained from their compounds by heating with carbon. For example, carbon cannot reduce the oxides of sodium, magnesium, calcium, aluminium, etc., to the respective metals. This is because these metals have more affinity for oxygen than carbon. These metals are obtained by electrolytic reduction. For example, sodium, magnesium and calcium are obtained by the electrolysis of their molten chlorides. The metals are deposited at the cathode (the negatively charged electrode), whereas, chlorine is liberated at the anode (the positively charged electrode). The reactions are –

At cathode $Na^+ + e- \rightarrow Na$
At anode $2Cl^- \rightarrow Cl_2 + 2e^-$

Similarly, aluminium is obtained by the electrolytic reduction of aluminium oxide.

Refining of Metals

The metals produced by various reduction processes described above are not very pure. They contain impurities, which must be removed to obtain pure metals. The most widely used method for refining impure metals is electrolytic refining.

Electrolytic Refining: Many metals, such as copper, zinc, tin, nickel, silver, gold, etc., are refined electrolytically. In this process, the impure metal is made the anode and a thin strip of pure metal is made the cathode. A solution of the metal salt is used as an electrolyte. On passing the current through the electrolyte, the pure metal from the anode dissolves into the electrolyte. An equivalent amount of pure metal from the electrolyte is deposited on the cathode. The soluble impurities go into the solution, whereas, the insoluble impurities settle down at the bottom of the anode and are known as *anode mud.*

Review questions

Objective questions:

Q1. Which metal is liquid at room temperature?

    a. Hg
    b. Br
    c. Cl
    d. None

Q2. Which non metal is liquid at room temperature?

    a. Hg
    b. Br
    c. Cl
    d. None

Q3. Which non metal is good conductor of electricity?

    e. Cu
    f. Fe
    g. Graphite
    h. Diamond

Q4. Which of the following can be cut by knife?

    a. Na
    b. Mg
    c. K
    d. All of the above

Q5. Calcination is used for?

    a. Sulphide ore
    b. Carbonate ore
    c. Oxide ore
    d. None of these

Q6. Which one is more reactive?

    a. Cu

    b. Pb

    c. Fe

    d. Zn

Q7. In thermite reaction the product is?

    a. Fe(l)

    b. Al_2O_3

    c. heat

    d. All of the above

Q8. In galvanization the layer of which metal is coated on iron or steel articles?

    a. Zn

    b. Mg

    c. K

    d. none of these

Q9. An alloy is a homogeneous mixture of?

    a. Two or more metals

    b. Metal and non metals

    c. Both a and b

    d. None of these

Q10. Al_2O_3 is?

    a. Metallic oxide

    b. Non metallic oxides

    c. Amphoteric oxide

    d. None

Q11. Which types of substances are effective in cleaning the vessels?

    a. sour

b. oily

c. sweets

d. none of these

Q12. In electrolytic refining the pure metal is taken as?

a. anode

b. cathode

c. anyone can be taken

d. none of these

Q13. Copper is used to make hot water tank because?

a. It is good conductor of heat

b. It is good conductor of electricity

c. malleability

d. ductility

Q14. Coins are manufactured from silver and copper because?

a. These are good conductor of heat

b. These are good conductor of electricity

c. Malleability

d. ductility

Q15. Non metallic oxides are?

a. acidic

b. neutral

c. either acidic or neutral

d. none of these

Q16. Alloy of Cu and Zn is?

a. bronze

b. solder

c. brass

d. amalgam

Q17. In amalgam which element must be present?

a. Hg
b. Pb
c. Fe
d. Zn

Q18.solder has?

a. High melting point
b. Low melting point
c. Moderate melting point
d. Extremely low melting point

Q19.cinebar is ore of?

a. Hg
b. Pb
c. Fe
d. None of these

Q20. Copper pyrite is ore of?

a. Cu
b. Pb
c. Fe
d. Zn

Match the following:

Q21.

A	B
amphoteric	CO_2
acidic	Al_2O_3
basic	NaOH
neutral	H_2O

Q22.

A	B
ionic	CH_4
co-valent	NaCl
roasting	$ZnCO_3$
calcination	ZnS

Q23.

A	B
solder	Zn & Cu
brass	Cu & Sn
bronze	Hg & Al
amalgam	Pb & Sn

short answer type questions:

Q24. What are physical differences between metal and non metals?

Q25. What are chemical differences between metal and non metals?

Q26. What are alloys? Give examples.

Q27. What is difference between ores and minerals?

Q28. What are amphoteric oxides?

Q29. What is thermite reaction?

Q30. What happens when zinc reacts with cold water and steam?

Q31. Which type of compound is formed when metal reacts with non metals?

Q32. What is difference between calcinations and roasting?

Q33. How does metals are extracted from its oxides?

Long answer type questions:

Q34. Give reasons
(a) Platinum, gold and silver are used to make jewellery.
(b) Sodium, potassium and lithium are stored under oil.
(c) Aluminium is a highly reactive metal, yet it is used to make utensils for cooking.
(d) Carbonate and sulphide ores are usually converted into oxides during the process of extraction.

Q35. Pratyush took sulphur powder on a spatula and heated it. He collected the gas evolved by inverting a test tube over it, as shown in figure below.
(a) What will be the action of gas on?
(i) Dry litmus paper?
(ii) Moist litmus paper?
(b) Write a balanced chemical equation for the reaction taking place.

Chapter 8

Carbon and its compounds

Introduction

Carbon is present in almost every matter around us. Maximum things around us on burning changed to the ashes (oxides of carbon). For example: the sugar we eat, cloths we wear and food items we eat all produces carbon on burning. It is such a versatile and Omni present element; it forms maximum no. of compounds in the world.

About carbon

As we have already studied its atomic no. is 6 and its atomic mass is 12 amu. It has 6 protons and 6 electrons in a C-12 carbon atom. During reaction whether it can take 4 electrons or loose 4 electrons in order to get a stable configuration. In first case 6 protons has to balance 10 electrons which seems unstable, in second case it requires much more energy to release 4 electrons. To overcome this problem carbon shares it electrons with other elements or same type carbon atoms.

The compounds which are formed from the combination of hydrogen and carbon are generally referred as hydrocarbons. However these compounds can have some other elements like sulphur, oxygen, nitrogen etc as a functional group. About functional group we will study in sub sequent sections.

Covalent bond

We have already studied in part 1, about covalent bonding. Here we will see that in covalent bonding the shared electrons is counted in both atoms to full fill the stable configuration. These bonding also can be expressed in the form of electron dot structure. Each electron is represented as a dot. The combination of two dots forms a single bond.

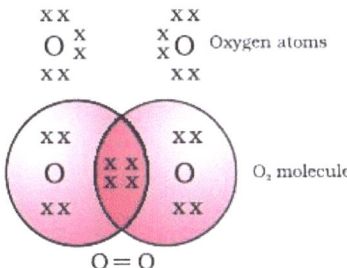

O = O

Versatile nature of carbon

There are two versatile property of carbon through which it forms infinite no. of carbon compounds. These are as follows:

1. Its four valency which provides a lot of branching.

2. Its property to link with same type of atoms generally known as *catenation*.

Saturated and unsaturated carbon compounds

Further hydrocarbons can be divided in two categories i.e. saturated and unsaturated. Saturated hydrocarbons contain single bonds while unsaturated contain double or triple bonds. There other differences also between them, for example: saturated hydrocarbons give blue flame on complete combustion while unsaturated give a yellow sooty flame. Saturated hydrocarbons undergo substitution reaction while unsaturated can't. Generally saturated are fairly inert in the presence of most of the reagents while unsaturated are not. Alkanes are saturated while alkenes and alkynes are unsaturated in nature.

Alkane

Alkanes have general formula C_nH_{2n+2}, where n refers to the no. of atoms. This formula shows that the no. hydrogen atom is two more than the double of the carbon atom. Meth, eth, prop but,pent, hex, hept, oct, non and dec signifies the no. of carbon atom 1, 2, 3, 4 5, 6, 7, 8, 9 and 10 respectively while –ane is joined to these root words to form the name of the compounds. The following table shows the names of first ten alkanes.

Chemical Formula	IUPAC Name
CH_4	Methane
C_2H_6	Ethane
C_3H_8	Propane
C_4H_{10}	Butane
C_5H_{12}	Pentane
C_6H_{14}	Hexane
C_7H_{16}	Heptane
C_8H_{18}	Octane
C_9H_{20}	Nonane
$C_{10}H_{22}$	Decane

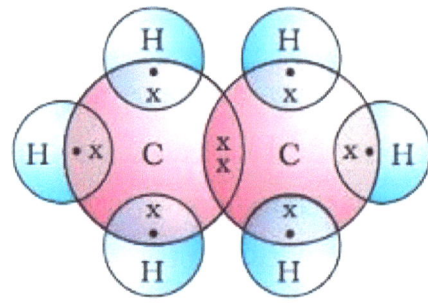

Alkene

Alkenes have general formula C_nH_{2n} , where n refers to the no. of atoms. This formula shows that the no. hydrogen atom is double of the carbon atom. In this compound there must be one double bond.

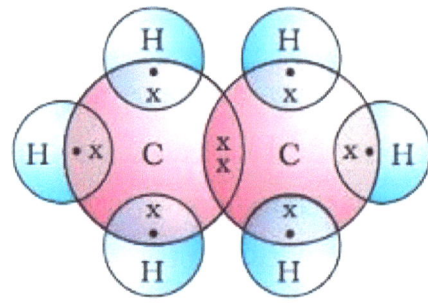

Alkyne

Alkynes have general formula C_nH_{2n-2}, where n refers to the no. of atoms. This formula shows that the no. hydrogen atom is two less than the double of the carbon atom. In this compound there must be one triple bond.

Butyne	C_4H_6	H $-$ C $\equiv$ C $-$ C $-$ C $-$ H (with H H on top and H H on bottom)
Pentyne	C_5H_8	H $-$ C $\equiv$ C $-$ C $-$ C $-$ C $-$ H (with H H H on top and H H H on bottom)

Functional groups

When one hydrogen atom is removed from the alkane, a new group is formed this is known as *alkyl group*. It is generally denoted by –R. Functional group is the atom or group of atoms which when attached with alkyl group can change its physical and chemical properties.

Halo- group

These are the halogen atoms, such as fluorine, chlorine, bromine, iodine etc.

1-chloroethane

H $-$ C $-$ C $-$ Cl (with H H on top and H H on bottom)

Alcohol group

This group consists of hydrogen and oxygen i.e. –OH.

H H
H—C—C—O—H
H H

Aldehyde group

Aldehyde group consists of C, H and O in the following fashion:

$$R — \overset{\overset{\displaystyle O}{||}}{C} —H$$

Ketone group

This group is —CO. in this group there is double bond between carbon and oxygen. And carbon is attached with at least two alkyl group.

$$CH_3 — \overset{\overset{\displaystyle O}{||}}{C} — CH_3$$

Acetone

Aldehyde and ketone

However this double bond between carbon and oxygen also present in the aldehyde group but it is ketone only in which carbon atom of functional group is attached to at least two alkyl groups.

Carboxylic acid

Carboxylic acid has the functional group –COOH. There is one double bond present between the carbon and oxygen, while there is single bond between hydrogen and oxygen.

$$
\begin{array}{c}
O \\
\parallel \\
C \\
R \diagup \quad \diagdown OH
\end{array}
$$

Nomenclature of carbon compounds

Naming a carbon compound can be done by the following method –
(i) Identify the number of carbon atoms in the compound. A compound having three carbon atoms would have the name propane.
(ii) In case a functional group is present, it is indicated in the name of the compound with either a prefix or a suffix (as given in Table 4.4).
(iii) If the name of the functional group is to be given as a suffix, the name of the carbon chain is modified by deleting the final 'e' and adding the appropriate suffix. For example, a three-carbon chain with a ketone group would be named in the following manner – Propane – 'e' = propan + 'one' = propanone.
(iv) If the carbon chain is unsaturated, then the final 'ane' in the name of the carbon chain is substituted by 'ene' or 'yne' as given in Table 4.4. For example, a three-carbon chain with a double bond would be called propene and if it has a triple bond, it would be called propyne.

Table 4.4 Nomenclature of functional groups

Functional group	Prefix/Suffix	Example	
1. Halogen	Prefix-chloro, bromo, etc.	H−C−C−C−Cl (with H H H above, H H H below)	Chloropropane
		H−C−C−C−Br (with H H H above, H H H below)	Bromopropane
2. Alcohol	Suffix - ol	H−C−C−C−OH (with H H H above, H H H below)	Propanol
3. Aldehyde	Suffix - al	H−C−C−C=O (with H H H above, H H below)	Propanal
4. Ketone	Suffix - one	H−C−C−C−H (with H H above, H O H below)	Propanone
5. Carboxylic acid	Suffix - oic acid	H−C−C−C−OH (with H H O above, H H below)	Propanoic acid
6. Double bond (alkenes)	Suffix - ene	H−C−C=C (with H H above, H below, H H on right)	Propene
7. Triple bond (alkynes)	Suffix - yne	H−C−C≡C−H (with H above, H below)	Propyne

Chemical properties of carbon compounds

Combustion: In combustion reaction the hydrocarbons on combustion produces heat and light.

(i) $C + O_2 \rightarrow CO_2$ + heat and light
(ii) $CH_4 + O2 \rightarrow CO_2 + H_2O$ + heat and light
(iii) $CH_3CH_2OH + O2 \rightarrow CO_2 + H_2O$ + heat and light

Oxidation: oxidation refers to the addition of oxygen. This reaction is generally used to convert alcohols to carboxylic groups. When we add oxygen to alcohol using an oxidizing agent like alkaline $KMnO_4$ or acidified $K_2Cr_2O_7$, alcohol changes to carboxylic acid.

$$CH_3 - CH_2OH \xrightarrow[\text{Or acidified } K_2Cr_2O_7 + \text{Heat}]{\text{Alkaline } KMnO_4 + \text{Heat}} CH_3COOH$$

Addition reaction: addition reaction is used to convert unsaturated hydrocarbon into saturated hydrocarbon. In this reaction a single bond is broken to associate with halogens. In many cases in place of halogen H_2 is used with nickel as a catalyst. it is commercially used to convert vegetable oil into animal fats.

Substitution reaction: Saturated hydrocarbons are fairly unreactive and are inert in the presence of most reagents. However, in the presence of sunlight, chlorine is added to hydrocarbons in a very fast reaction. Chlorine can replace the hydrogen atoms one by one. It is called a substitution reaction because one type of atom or a group of atoms takes the place of another. A number of products are usually formed with the higher homologues of alkanes.

$CH_4 + Cl_2 \rightarrow CH_3Cl + HCl$ (in the presence of sunlight)

Some important carbon compounds

Ethanol

Physical properties

Ethanol is a liquid at room temperature (refer to Table 4.1 for the melting and boiling points of ethanol). Ethanol is commonly called alcohol and is the active ingredient of all alcoholic drinks. In addition, because it is a good solvent, it is also used in medicines such as tincture iodine, cough syrups, and many tonics. Ethanol is also soluble in water in all proportions. Consumption of small quantities of dilute ethanol causes drunkenness. Even though this practice is condemned, it is a socially widespread practice. However, intake of even a small quantity of pure ethanol (called absolute alcohol) can be lethal. Also, long-term consumption of alcohol leads to many health problems.

Chemical properties

(i) Reaction with sodium –
$$2Na + 2CH_3CH_2OH \rightarrow 2CH_3CH_2O^-Na^+ + H_2$$
$$\text{(Sodium ethoxide)}$$
Alcohols react with sodium leading to the evolution of hydrogen. With ethanol, the other product is sodium ethoxide.

(ii) Reaction to give unsaturated hydrocarbon: Heating ethanol at 443 K with excess concentrated sulphuric acid results in the dehydration of ethanol to give ethene –

$$CH_3 CH_2OH \rightarrow CH_2=CH_2 + H_2O$$

The concentrated sulphuric acid can be regarded as a dehydrating agent which removes water from ethanol.

Ethanoic acid

Physical properties

Ethanoic acid is commonly called acetic acid and belongs to a group of acids called carboxylic acids. 5-8% solution of acetic acid in water is called vinegar and is used widely as a preservative in pickles. The melting point of pure ethanoic acid is 290 K and hence it often freezes during winter in cold climates. This gave rise to its name glacial acetic acid. The groups of organic compounds called carboxylic acids are obviously characterized by a special acidity. However, unlike mineral acids like HCl, which are completely ionized, carboxylic acids are weak acids.

Chemical properties

(i) *Esterification reaction:* Esters are most commonly formed by reaction of an acid and an alcohol. Ethanoic acid reacts with absolute ethanol in the presence of an acid catalyst to give an ester –

$$CH_3COOH + CH_3CH_2OH \rightarrow CH_3\text{-}CO\text{-}CH_2CH_3$$
(ethanoic acid)　(alcohol)　acid　(Ester)

Esters are sweet-smelling substances. These are used in making perfumes and as flavouring agents. Esters react in the presence of an acid or a base to give back the alcohol and carboxylic acid. This reaction is known as saponification because it is used in the preparation of soap.

Formation of ester

$$CH_3COOC_2H_5 \: C \: H \: OH \rightarrow C_2H_5OH + CH_3COOH$$
$$(NaOH)$$

(ii) *Reaction with a base:* Like mineral acids, ethanoic acid reacts with a base such as sodium hydroxide to give a salt (sodium ethanoate or commonly called sodium acetate) and water:

$$NaOH + CH_3COOH \rightarrow CH_3COONa + H_2O$$

(iii) *Reaction with carbonates and hydrogen carbonates:* Ethanoic acid reacts with carbonates and hydrogen carbonates to give rise to a salt, carbon dioxide and water. The salt produced is commonly called sodium acetate.

$$2CH_3COOH + Na_2CO_3 \rightarrow 2CH_3COONa + H_2O + CO_2$$
$$CH_3COOH + NaHCO_3 \rightarrow CH_3COONa + H_2O + CO_2$$

Soaps and detergents

Detergents are ammonium or sulphonate salts of long chain carboxylic acids while soaps are sodium salt of carboxylic acid.
When calcium or magnesium salts are dissolved in water it becomes hard water and can't be useful for soap due to formation of scum an insoluble substance. However detergent can be used in hard water because charged ends of detergent do not form scum.

The following table shows the difference between the soaps and detergent.

SOAPS	DETERGENTS
They contain sodium carboxylate (COONa) group.	They contain sodium sulphonate (SO_3Na) group.
They are not suitable for washing with hard water.	They are suitable for both hard and soft water.
They have relatively weak cleansing action.	They have strong cleansing action.
They are biodegradable.	Most of them are non-biodegradable.

Cleaning action of soaps

The action of soaps and detergents is based on the presence of both hydrophobic and hydrophilic groups in the molecule and this helps to emulsify the oily dirt and hence its removal.

When detergent or soaps are dissolved in water they produce many molecules which have hydrophobic and hydrophilic ends. Hydrophobic end repel water molecule and get attached to the dirt while hydrophilic does its opposite. In this process all hydrophobic parts of molecules get attached to the dirt and pulled by water molecules, this unique formation is named *micelle*. When we wash out water dirt is removed and the cloth gets cleaned.

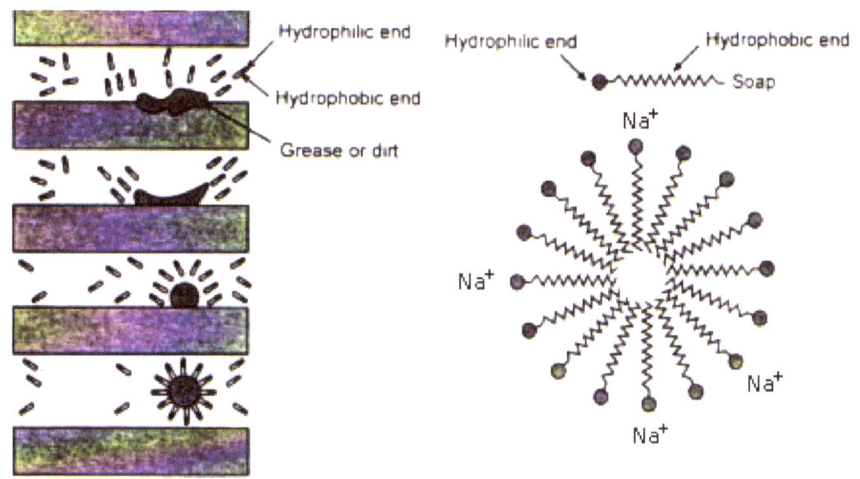

Review questions

OBJECTIVE QUESTIONS:

Q1. The % of carbon in earth crust is?

    a. 0.002

    b. 0.02

    c. 20

    d. 0.2

Q2. Carbon has valency?

    a. 1

    b. 2

    c. 3

    d. 4

Q3. Catenation is found in?

    a. C

    b. O

    c. F

    d. None of these

Q4. Bonding in carbon is?

a. covalent

 b. ionic

c.co-ordinate

d. none of these

Q5. A saturated hydrocarbon has?

    a. Single bond

    b. Double bond

c. Triple bond

d. None of these

Q6. Formula of benzene is?

    a. C_2H_6

    b. C_6H_6

    c. C_5H_6

    d. None of these

Q7. Homologous series differ by?

    a. CH_2 unit

    b. 14 amu by mass

    c. Both a and b

    d. None of these

Q8. $CH_3CH_2CH_2OH$ is?

    a. alcohol

    b. ketone

    c. carboxylic acid

    d. None of these

Q9. In oxidation reaction the oxidising agent is?

    a. Alkaline $KMnO_4$

    b. Acidified $K_2Cr_2O_7$

    c. Both a and b

    d. None of these

Q10. In addition reaction the catalyst used is?

    a. H_2

    b. Acidified $K_2Cr_2O_7$

    c. nickel

    d. None of these

Q11. The conversion of ethanol to ethanoic acid is?

    a. Oxidation reaction

    b. Addition reaction

    c. Substituition reaction

    d. None of these

Q12. Which is commonly known as alcohol?

    a. ethanol

    b. ethanoic acid

    c. methanol

    d. None of these

Q13. Vinegar is formed from?

    a. ethanol

    b. ethanoic acid

    c. methanol

    d. None of these

Q14. In esterification the catalyst used is?

    a. base

    b. acid

    c. salt

    d. None of these

Q15. Saponification is the reverse process of?

    a. carbocation

    b. esterification

    c. carbocation

    d. None of these

Q16. Ionic end of soap dissolves in?

    a. water

b. oil

c. acid

d. base

Q17. Shampoos are?

a. soap

b. detergent

c. salts

d. acids

Q.18 which of the following can go under addition reaction?

a. C_2H_6

b. C_6H_6

c. C_5H_6

d. None of these

Q19. CH_3COOCH_3 is?

a. soap

b. acid

c. base

d. ester

Q20. Which of the following causes hardness of water?

a. Calcium salt

b. Magnesium salt

c. Both a and b

d. None of these

Match the following:

Q21.

A	B
C-60	saturated
ethene	allotrophes
methane	unsaturated
carbon	catenation

Q22.

A	B
-OH	carboxyllic acid
-CHO	ketone
-CO-	aldehyde
-COOH	alcohol

Q23.

A	B
nickel	substituition reaction
sun light	addition reaction
alkaline $KMnO_4$	oxidation reaction
hot conc. H_2SO_4	formation of ethene

Short answer type questions:

Q24. What are the versatile properties of carbon?

Q25. Write down differences between saturated and unsaturated hydrocarbons?

Q26. What is homologous series?

Q27. What is addition reaction?

Q28. What is substitution reaction?

Q29. Write down physical properties of ethanol?

Q30. Write down physical properties of ethanoic acid?

Q31. Write down chemical properties of ethanol?

Q32. Write down chemical properties of ethanoic acid?

Long answer type questions:

Q33. Draw the structures for the following compounds.
    (i) Ethanoic acid   (ii) Bromopentane*
    (iii) Butanone   (iv Hexanal.
    *Are structural isomers possible for bromopentane?

Q34. Explain the cleaning action of soap?

Q35. Explain the phenomenon of esterification and saponification in detail with examples.

Chapter 9

Periodic classification of elements

Introduction

Why classification is important?

Classification provides us a easy way to understand the properties and access to the particular element. We can take a simple example; in a mall the items pertaining to same purposes are kept one side. If all items present in the mall are spread and mixed, what will happen? It will create a huge disturbance to customer as well as managing team. If it is classified, it is quite easy for selling as well as purchasing.

If all elements are arranged according to some particular properties, it will be quite easier to select a particular element for a particular purpose. Suppose any engineer wants to form the body part of an aero plane. Then first, he will be attempted to ignore liquids and gases. Then he will try to choose light metals which have high tensile strength. Further he may be interested in other properties too. All these comparisons can be done only when all elements are arranged in a particular way. But before 200 years when classification starts, it was a great quest, how to start it? There came thousands of ways and tables some were discarded; some were improved and gave a platform to next one. In this chapter we will see only some of the classification which proved to the base for modern periodic classification of elements.

In this chapter our strategy will be to have a look at the basic rule for a particular classification then its merits and demerits. We will also deal with modern periodic table in details.

Dobereiner's triads

In the year 1817, Johann Wolfgang Döbereiner, a German chemist, arranged elements in the group of three named them as triads. According to dobereiner

When these three elements are arranged in increasing order of their atomic weight the middle weight is average of remaining two. Following table shows the examples.

Elements	Atomic Mass	Arthmetic Mean
Lithium Sodium Potassium	7 23 39	$\dfrac{7+39}{2}=23$
Chlorine Bromine Iodine	35.5 80 126.5	$\dfrac{35.5 + 126.5}{2}=81$
Calcium Strontium Barium	40 87 137	$\dfrac{137+40}{2}=88$

Demerits

Dobereiner could arrange only three groups.

Newland's octaves

Doberiener's triads encouraged many researchers to search periodicity in elements according to the increasing order of mass. Newlands took the hydrogen as lightest element and thorium as 56th and last element. On arranging these elements in increasing order of their atomic mass, Newlands found that every eighth element shows similar properties. He proposed a word "the octaves law" for this periodicity.

Demerits of Newlands octaves

1. Newlands law of octaves was applicable to only light elements. It was well up to calcium element.
2. Newland announces that there are only 56 elements in the nature and no further elements will be discovered. But soon it was realized that it is not true.
3. Some elements were put in same slot like cobalt and nickel and both of these were in the column of fluorine which has different properties. Iron which resembles properties same as cobalt was kept far away from it.

Newlands' Octaves

H	Li	Be	B	C	N	O
F	Na	Mg	Al	Si	P	S
Cl	K	Ca	Cr	Ti	Mn	Fe
Co, Ni	Cu	Zn	Y	In	As	Se
Br	Rb	Sr	Ce, La	Zr	Di, Mo	Ro, Ru
Pd	Ag	Cd	U	Sn	Sb	I
Te	Cs	Ba, V	Ta	W	Nb	Au
Pt, Ir	Os	Hg	Tl	Pb	Bi	Th

Mendeleev's periodic table

When Mendeleev started preparing the periodic table there were 63 elements. Mendeleev investigated the periodic function of masses as well as physical and chemical properties. He formed oxides and hydrides of each element and kept similar in a same group. Then he stated the law of periodic table which states that which states that 'the properties of elements are the periodic function of their atomic masses'.

Achievements of Mendeleev's periodic table:

Mendeleev's left some gaps in the periodic table. He took it as a merit and explained that some elements will be discovered in the future and will be kept at these positions. These elements were scandium, gallium, germanium and were respectively named *as **Eka–boron, Eka–aluminium and Eka–silicon***. Here eka (a snskrit word) means one. Eka word was prefix to the element preceding in the same group. Predictions of properties were almost same as these elements have.

His periodic table was so flexible that when inert gases were discovered they got their places easily.

Limitations of Mendeleev's periodic table:

Mendeleev was ***unable to give a fix position for hydrogen***. His periodic table was based on physical and chemical properties and hydrogen has these properties similar to metal as well as non metals.

As periodic table was arranged in according to the increasing order of their atomic mass there was ***no any place for isotopes***.

Legend: Dobereiner's triads | Known to Mendeleev | Unknown to Mendeleev

							H 1.01				
He 4.00	Li 6.94	Be 9.01	B 10.8	C 12.0	N 14.0	O 16.0	F 19.0				
Ne 20.2	Na 23.0	Mg 24.3	Al 27.0	Si 28.1	P 31.0	S 32.1	Cl 35.5				
Ar 40.0	K 39.1	Ca 40.1	Sc 45.0	Ti 47.9	V 50.9	Cr 52.0	Mn 54.9	Fe 55.9	Co 58.9	Ni 58.7	
	Cu 63.5	Zn 65.4	Ga 69.7	Ge 72.6	As 74.9	Se 79.0	Br 79.9				
Kr 83.8	Rb 85.5	Sr 87.6	Y 88.9	Zr 91.2	Nb 92.9	Mo 95.9	Tc (99)	Ru 101	Rh 103	Pd 106	
	Ag 108	Cd 112	In 115	Sn 119	Sb 122	Te 128	I 127				
Xe 131	Ce 133	Ba 137	La 139	Hf 179	Ta 181	W 184	Re 180	Os 194	Ir 192	Pt 195	
	Au 197	Hg 201	Tl 204	Pb 207	Bi 209	Po (210)	At (210)				
Rn (222)	Fr (223)	Ra (226)	Ac (227)	Th 232	Pa (231)	U 238					

The modern periodic table

This periodic table was prepared by Henry Moseley in 1913. In this table Moseley took the atomic no. as the as a fundamental property rather than atomic mass. This concept solved many limitations of Mendeleev's periodic table. This law can be stated as follows:

'Properties of elements are a periodic function of their atomic number.'

Position of elements in modern periodic table

In modern periodic table, there are 7 horizontal rows known as periods and 18 vertical columns known as groups.

The elements having same chemical and physical properties are arranged in the same group. However in periods a gradation or slight change in physical and chemical properties is observed.

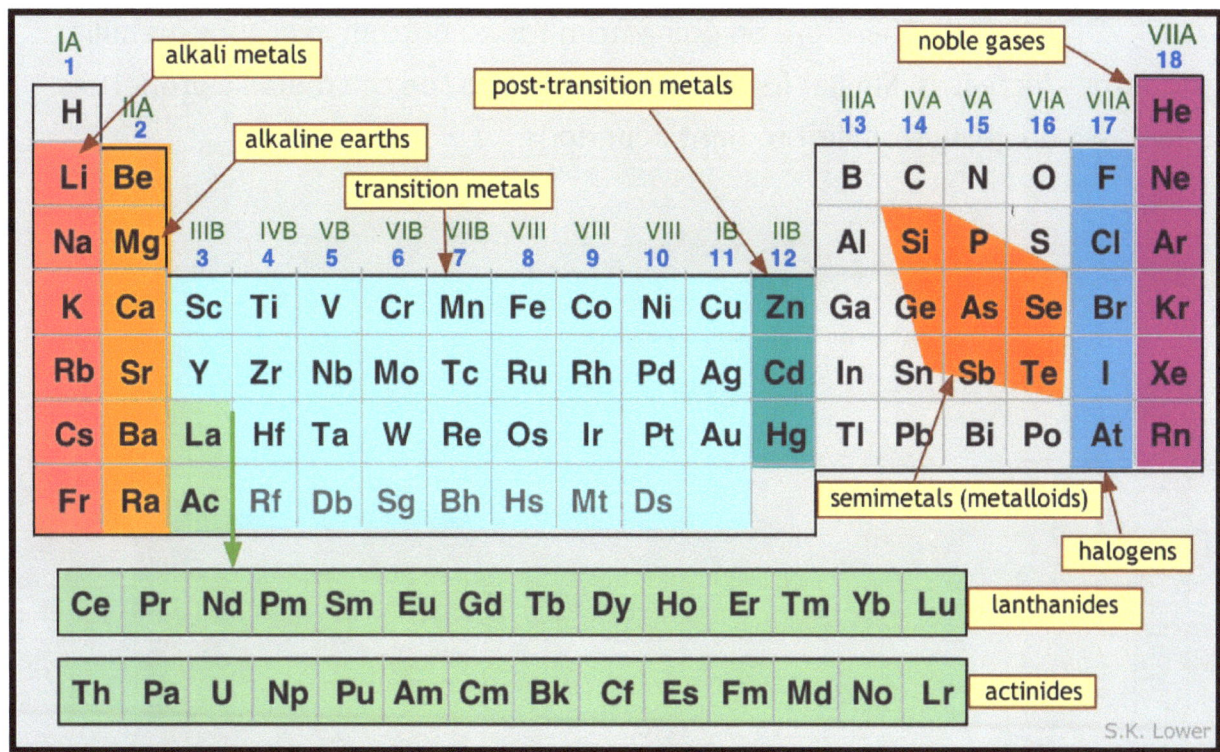

Trends in the modern periodic table

Valency: in a group the valency remains unchanged as there are same no. of electrons in outer most shell of atom. When we move from left to right in a period the valency first increases and then decreases. We have already seen how to calculate valency in "basic chemistry 1" if we take the second period of modern periodic table the elements Li, Be, B, C, N, O, F, and Ne we get the valencies as 1, 2, 3, 4, 3, 2, 1, 0 respectively.

Atomic size: in group as we go from top to bottom, due to increase in no. of orbital atomic size increases. When we move in periods from left to right the no. of orbitals remain same but no. of protons increases which attracts electrons and size shrinks i.e. atomic size decreases.

Metallic and non metallic properties: as we have seen in above section that when we go from top to bottom in groups the atomic size Increases and hence the electrons in the outer most shells are more free to move, this contribute to metallic properties.

Therefore on going from top to bottom in groups metallic properties increases. Similar logic can be applied to see the metallic properties decreases on going from left to right in periods.

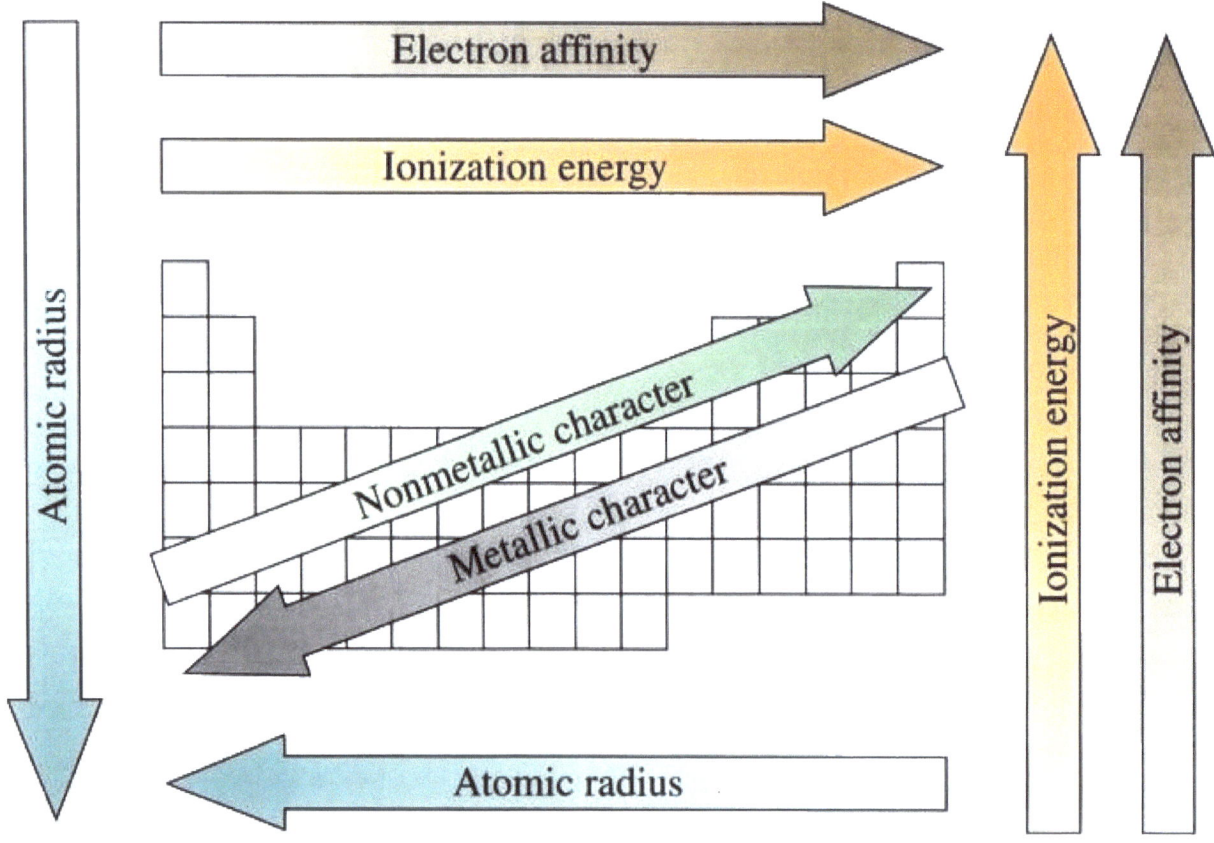

Review questions

Objective questions:

Q1. Doberiener arranged the elements in the form of?

    a. triads
    b. octaves
    c. tetrads
    d. none

Q2. Newlands arranged how many elements?

    a. 56
    b. 63
    c. 114
    d. 118

Q3. Mendeleev left gape for which element in his periodic table?

    a. scandium
    b. gallium
    c. germenium
    d. all of these

Q4. Mendeleev used basic concept of?

    a. Physical properties
    b. Chemical properties
    c. Atomic masses
    d. all of these

Q5. Who prepared modern periodic table?

    a. newland
    b. doberiener
    c. mendeleev
    d. moselley

Q6. How many periods are there in modern periodic table?

    a. 18
    b. 7
    c. 20
    d. None of these

Q7. What is the valency of magnesium?

    a. 1
    b. 2
    c. 3
    d. 4

Q8. How many groups are there in modern periodic table?

    a. 18
    b. 7
    c. 20
    d. None of these

Q9. In periods going right from left, the valency?

    a. increases
    b. decreases
    c. first increases then decreases
    d. None of these

Q10. In groups going from top to bottom, the valency?

    a. increases
    b. decreases
    c. first increases then decreases
    d. remain same.

Q11. In periods going right from left, atomic size?

    a. increases

b. decreases

c. first increases then decreases

d. None of these

Q12. In groups going from top to bottom, the atomic size?

a. increases

b. decreases

c. first increases then decreases

d. remain same.

Q13. In groups going from top to bottom, the metallic properties?

a. increases

b. decreases

c. first decreases then increases

d. remain same.

Q14. In periods going right from left, the non mettalic properties?

a. increases

b. decreases

c. first increases then decreases

d. None of these

Match the followings

Q15.

A	B
Law of octaves	moselley
triads	mendeleev
gaps in table	newlands
place for isotopes	doberiener

Q16.
	A	B
	56 elements	moselley
	63 elements	mendeleev
	9 elements	newlands
	118 elements	doberiener

Long answers type questions:

Q17. Explain the trends in the periodic table for

a. Valency
b. Atomic size
c. Mettalic and non metallic properties

Q18. Which element has
(a) Two shells, both of which are completely filled with electrons?
(b) The electronic configuration 2, 8, 2?
(c) A total of three shells, with four electrons in its valence shell?
(d) A total of two shells, with three electrons in its valence shell?
(e) twice as many electrons in its second shell as in its first shell?

Q19. (a) What property do all elements in the same column of the Periodic Table as boron have in common?
(b) What property does all elements in the same column of the Periodic Table as Fluorine has in common?

Q20. The position of three elements A, B and C in the Periodic Table are shown below –

Group 16	Group 17
-	-
-	A
-	-
B	C

(a) State whether A is a metal or non-metal.

(b) State whether C is more reactive or less reactive than A.

(c) Will C be larger or smaller in size than B?

(d) Which type of ion, cation or anion, will be formed by element A?

Answers to the review questions

Chapter 5

Chemical Reactions and Chemical Equations

Q1.a

Q2.c

Q3.a

Q4.c

Q5.c

Q6.a

Q7.c

Q8.b

Q9.a

Q10.a

Q11.d

Q12.b

Q13.a

Q14.a

Q15.b

Q16.d

Q17.c

Q18.a

Q19.b

Q20.b

Q21.

a---b

b---a

c---d

d---c

Q22.

a---b

b---a

c---d

d---c

Q23.

a---b

b---a

c---d

d---c

Chapter 6

Acid, Bases and Salts

Q1.a

Q2.a

Q3.c

Q4.a

Q5.b

Q6.a

Q7.b

Q8.a

Q9.c

Q10.d

Q11.c

Q12.c

Q13.b

Q14.b

Q15.a

Q16.b

Q17.d

Q18.a

Q19.b

Q20.c

Q21.

a---c

b---d

c---b

d---a

Q22.

a---c

b---e

c---d

d---b

e---a

Q23.

a---c

b---d

c---b

d---a

Chapter 7

Metals and Non Metals

Q1.a

Q2.b

Q3.c

Q4.d

Q5.b

Q6.b

Q7.d

Q8.a

Q9.c

Q10.c

Q11.a

Q12.b

Q13.a

Q14.c

Q15.c

Q16.c

Q17.a

Q18.b

Q19.a

Q20.c

Q21.

a---b

b---a

c---c

d---d

Q22.

a---b

b---a

c---d

d---c

Q23.

a---d

b---a

c---b

d---c

Chapter 8

Carbon Compounds

Q1.b

Q2.d

Q3.a

Q4.a

Q5.a

Q6.b

Q7.c

Q8.a

Q9.c

Q10.c

Q11.a

Q12.a

Q13.b

Q14.b

Q15.b

Q16.a

Q17.b

Q18.a

Q19.d

Q20.c

Q21.

a---b

b---c

c---a

d---d

Q22.

a---d

b---c

c---b

d---a

Q23.

a---b

b---a

c---c

d---d

Chapter 9

Periodic Classification Of Elements

Q1.a

Q2.a

Q3.d

Q4.d

Q5.d

Q6.b

Q7.b

Q8.a

Q9.c

Q10.d

Q11.b

Q12.a

Q13.a

Q14.a

Q15.

a---c

b---d

c---b

d---a

Q16.

a---c

b---b

c---d

d---a

NOTES